WHO WILL SPEAK FOR EARTH?

ALSO FROM
PATH OF POTENTIAL:

·Becoming. *Right for the Heart...*
Good for the Whole

·Gifts of the Spirit. *Experiencing Death and Loss
from the Perspective of Potential*

and numerous writings on the website:
www.pathofpotential.org

To order books, call:

Melody Fraser
The Mail Suite
1-800-818-6177
970-241-8973

WHO WILL SPEAK FOR EARTH?

Reflections on Securing Energy from a Life of the Whole Perspective

Terry P. Anderson
Sandra Maslow Smith

Path of Potential

WHO WILL SPEAK FOR EARTH?

PHOTOGRAPHY
Carol Anderson
Bill Clark
Candi Clark
Sandra Maslow Smith

COVER AND BOOK DESIGN
Candi Clark

GRAPHICS AND PRINTING
Candi Clark
Sunburst Graphics & Printing, Inc.
Grand Junction, CO 81503 USA

STYLISTIC EDITOR
Paige Gengenbach

PUBLISHER
Path of Potential
P.O. Box 4058
Grand Junction, CO 81502 USA
www.pathofpotential.org

AUTHORS
Terry P. Anderson
Sandra Maslow Smith

First Printing - 2006
Printed in the United States of America
30% Post Consumer Waste Fiber,
Acid Free, Archival Quality

Path of Potential is a trademark of TS Potential, LLC.

ISBN-10: 0-9760139-3-2
ISBN-13: 978-0-9760139-3-8

***This book is dedicated to
Helen Anderson***
1913-2005

*A mother to many...
A wise woman,
A woman of wisdom,
A woman of everlasting beauty,
A woman through whom much love
enters into the working of the world.*

"Who will Speak for Earth?" is a book of reflections written in the language of intuition. As such it contains neither conclusions nor opinions commonly gained by analysis of facts and data, but rather images of truth that both seek and lend themselves to further deepening and increasing wholeness.

CONTENTS

Prologue... Reflecting on the Question 9

Opening Our Hearts to Wisdom 13

Understanding 17

Earth Working 21

Living Philosophy of Potential 25

Ethics, Ethical Guidelines 29

    ·No occlusion to the "tuning" of the planetary life
        energy field. 33

    ·On entering a territory, seek to understand and honor,
        rather than interfere with, the spiritual culture and
        other essential life processes. 39

    ·Honor and understand the working of the life of the
        whole. 43

    ·Pursue technology and approaches that enable
        community self determination and environmental
        amelioration. 49

    ·Create processes that engender understanding and
        discourage divisiveness. 55

Community Principles 59

Scientific Theories and Knowledge 67

Community Dialogue 73

Intentional Manifestation 79

Upward Progressions 87

Community Integration 93

A Final Reflection...Gathering Our Community 99

Prologue

REFLECTING ON THE QUESTION

Who will speak for earth? Reflecting on this question, we notice...
A person cannot; a community perhaps, through time and effort; a people for sure: a people of the land... a people of earth... a people who see and understand themselves to be part of the whole – the whole of life... a people who are

clear that they are not separate from life – they are not outside of life and the working of life on earth – but rather integral members of earth's life community.

We know from experience…
To speak for another, if we seek to be authentic in our expression, is a very demanding task, a task that requires much reflection, a bit of what we often think of as "soul searching," and without doubt, an opening of our heart to the experience, the wisdom, the character of the heart of the one for whom we wish to speak. For it is the heart open to wisdom and enriched with purpose that lies behind that which we experience as the essence of the person… the force behind that which is manifested… the expression of life – their way of living, of going about life, and their way of working, of going about their life's work.

To speak for earth then, is to speak from the heart…
It is not a process that begins with analysis, but rather one that begins with wholeness – with intuition, the eye of the heart. To speak for earth is speaking with awe for that which has been created, an awe visible to a mother's eye, an eye that can see beauty in all… in all the manifestations of earth's living systems – the trees, the flowers, the cactus, the mountains, the desert, the forests, the oceans, etc. To speak for earth is to speak with deep appreciation for the tireless and the forgiving nature of mother earth's efforts to sustain the presence and working of life on earth… to speak for earth is to hear her crying out – crying out for a shift in our way of being and doing.

To speak for earth is not in dullness…
To speak for earth is not in dullness, the dullness of the sameness of voice, nor is it in voices of anger, but rather in the joyous uniqueness of voices, uniquenesses expressed

with increasing oneness... a oneness that transcends the divisiveness, the fractionation and the separation that are the antithesis of the systemicness of earth's living system character. To speak for earth is to accept the truth present in the essence of the way Helen Anderson lived her life: with love, and only through love, all things are possible. Love must be in the process of our coming together to develop the capacity to speak for earth.

Reflecting further...
We notice that the requirement – perhaps more accurately the urge or call – to speak for another often emerges during moments or times of significance, be they special moments of joy and goodness, or times when we are faced with an issue of significance... an issue like that of securing energy... an issue clearly facing ourselves, not only ourselves, but the whole of life as well... an issue that requires we develop wholeness... wholeness of understanding... wholeness of the truth of our being members of the community of life on earth... wholeness in our transcending our being human-centered, to our embracing and becoming life-of-the-whole-centered, and thereby seeing and holding the perspective of our human potential unfolding and contributing to – doing our part in – the advancing of the life of the whole.

Completing our reflection...
Reflecting on the way of bees – fellow creatures of life – we see that among some bees, among some hives, it is common to send out a number of their members in search of nectar (flowers, clover, etc.), nectar being essential to sustaining the bees' life processes, their life itself. After discovering a "patch" of nectar, the bee returns to the hive, at which time other bees of the hive engage the returning member in a process often thought of as a dance... a dance involving the many... an intricate, an elaborate, an intentional

pattern. It is through this dance that understanding and knowledge are generated – a mind is created – that not only exceeds the "brain power" of a single bee, but also exceeds the understanding and knowledge brought in by the returning bee.

It is in the spirit of the returning bee that this book is written. It is hoped that the reflections found within will have sufficient "nectar of truth" to be worthy of a dance... or two... or perhaps three... among the "hives" and "hive" of humanity. For as Helen Anderson understood that it is only through love that all things become possible, we in a similar way, have come to understand that the truth – the truth of wisdom - is as essential to our well being, our humanness and our becoming fully and truly human, as nectar is to the bees. It is through the truth of wisdom that we will begin to be able to see and accept our work and role in the world, on this planet earth, in much the same way bees understand pollination to be their work, their role... herein, we believe, lies much hope.

Chapter One

OPENING OUR HEARTS TO WISDOM

A Time for Grassroots Process...
There are times in the life cycles of America where we have to regenerate the spirit, essence and energy of our country and our country's people so that the direction we take and the path we

choose reflects who we truly are and what we are earnestly and honestly trying to become... not only to and for ourselves, but for the rest of the world as well. Such times call for grassroots process – the process by which we the people come together to reflect, to dialogue, to image and to imagine… to be and become clear as to what is truly at stake and as to what choices to make – wise choices, choices grounded in reality, choices that are right and good for each and all. Today is such a time – a grassroots time – for America… and beyond.

A Need for Understanding...
Today, as a country, we are actively engaged in the pursuit of securing energy. As we reflect on securing energy - as the image of that process begins to develop and unfold for us - we notice that the notion of securing energy not only encompasses the physical energies required for locomotion, heating and cooling, but also fully embraces the vitalizing and spirit-lifting energies present throughout earth... energies like those available from experiencing a walk on a beach at sunrise, the freshness of food, the silence and beauty of the wilderness, and a picturesque and vibrant harbor. This securing energy pursuit is calling upon the involvement - the grassroots involvement - of we the people… grassroots not in the sense that we need to resist or deny this necessity, but rather because we need to develop within and among us an understanding from which can emerge wise choices and right and good decisions. Given the scope and magnitude – the obvious and significant affect our actions will have on earth, the life processes of earth, and the whole of life, ourselves included – it behooves us and requires us to be diligent and deliberate in our approach.

A Call for True Intuition...
Securing energy is an issue that can be left neither solely to the "experts," nor solely to the political process. It requires a

wholistic and systemic understanding that itself calls upon the perspective of many. This critical issue of our time also demands the inclusion and use of intuition – true intuition, that which we call upon and bring to bear when we seek to see wholeness, essence, systemic relatedness, and intended working... the very elements present to wisdom.

A Need for a New Frame of Reference...
More and more often, among more and more of us, we find we are seriously questioning the path we are on. Intuitively, in our heart of hearts, we are beginning to see and understand the necessity for a new path – a path of increasing, rather than diminishing, hope. Such a path requires that we develop our thinking and actions from a frame of reference that enables us to see the world and ourselves in it, not from the perspective of segmented problems and issues, but rather from the perspective and the approach that embraces the whole of life, and the life of the whole. And so we have before us a real task of building for ourselves a frame of reference – an orderly and whole thought base from which to live… a frame of reference we can call upon and reflect against as we make our way into a world different from the one we have commonly experienced. This new thought base seeks not to discard essential truths, but rather to make them real – truly relevant – in the dynamic and emerging world we are being called upon to shape and create. We sense that in the absence of a real, practical, working frame of reference – a frame of reference that ongoingly aids in seeing essential truths – we will lose our opportunity to be and become… and instead be caught up in, defined by, and swept away by unknowing conformance to past likes, dislikes, habits, fads, attachments, etc.

A frame of reference that is real and relevant to the dynamics and drama of the time in which we are living, and one that reflects our inherent urge or call to become, is a living philosophy. A living philosophy is one that we can live from and be disciplined about, that acknowledges the reality of our current situation, and that enables the realization of the open-ended potential of each and all... a philosophy that is personal, yet one that operates in service of the larger wholes within which we live – our communities, society, the whole of life. A living philosophy both comes from potential and has the potential to bring about a wholeness and unity of family, community and humanity previously unimaginable. As such, it is a people's philosophy because it flows through and among the people through reflection and dialogue, guided by purpose and principles, acted on and carried out by heartfelt roles and expressions.

What follows here is written to provide food for reflection and dialogue – the nature of reflection and dialogue that contributes to creating our community living philosophy of potential... a philosophy that encompasses one's own life, as well as the whole of life on this earth.

Chapter Two
UNDERSTANDING

It is understanding we seek when we wish to live with integrity... to live with an increasing sense of wholeness... wholeness within ourselves, within our families, within our communities.

It is through understanding that partial or selected truths begin to give way to whole truths… whole truths that bring to the fore, simplicity… simplicity and authenticity.

It is through understanding that purposes become clarified… and inner conflicts give way toward an increasing sense of oneness… a oneness and growing sense of completeness.

It is through understanding that we become more able to live a life we believe in… a life in harmony with our heart of hearts, a wiser life, a truer life, a truer path.

It is through "seeing" from the perspective of potential that we develop understanding – true understanding – understanding that can become for us our living philosophy of potential.

CYCLE OF UNDERSTANDING RELATIVE TO LIFE ON EARTH

Chapter Three
EARTH WORKING

Earth works as a system - a living system. As such she operates under the same reality affecting all living systems: if you touch one element, you touch them all. Any action we take towards a particular element affects the health and the working of all the elements… and subsequently the health and well being of the whole of which the systems serve. Air, water, soil and sun are systemic elements of the living processes of earth. Actions that affect the intended and orderly working of any or all of these elements affect all of life, and the whole of life.

Earth organizes herself in planetary energy fields – life energy fields. If we reflect on the desert, the shore, the forest land, the mountains, the ocean, the prairie, etc., we can begin to bring some depth and substance to the image of the planetary life energy fields of earth. We notice the particular and the unique character of each - the ability for ourselves to tune into and to be tuned in a particular way by a particular life energy field. Further noticing provides an imagery of the nature and type of flora and fauna that both inhabit and contribute to the life energy field, a nature and type that we see flowing into our nature as images of people of the desert, people of the shore, people of the plains, etc., come to the fore.

To sustain the vitality, the viability and the unfolding of earthly life, earth engages in a multitude of processes... processes we commonly attribute to mother nature... processes that nourish, replenish, and regenerate the ongoingness of life. Some of these processes - softly falling rain, the warming feel of the sun, gentle cooling breezes - are invitingly pleasant to us. Some others, such as fires, floods, hurricanes and earthquakes, are less pleasant, perhaps even frightening, but nevertheless inherent to the healthy working of the life processes of earth... and whereas our actions can greatly influence the severity and perhaps the frequency and duration of these regenerative processes of earth, it would seem unwise to work to interfere with their intentional and orderly workings.

Chapter Four
Living Philosophy of Potential

The whole universe, including this place we call earth, came into existence for a reason, for a purpose… and it is moving toward the fulfillment of that – toward the realization of its potential. This becoming, this unfolding, includes not only the physical/material perspective we commonly think of as evolution, but also the advancement of being, and the manifestation of will as well.

Earth is a means for life to enter into the working of and the unfolding of the universe. Purpose, wholeness, oneness and systemic relatedness are essential characteristics of life and living beings. We as living human beings share these characteristics common to all of life… and like all of life, we struggle to survive and strive to become - to realize our potential, to fulfill the intent of our design.

As living human beings, as an aspect of our design, we can consciously and conscientiously act in accord with an understanding of the work and working of the whole – we can acknowledge and honor not only the particular essential processes of humanity, but also the essential life processes of earth and those of the community of life – a community of which we are members. Through true intuition – the eye of the heart – we can gain access to the truth and wisdom a living philosophy of potential would require... the truth and wisdom in regard to wholeness, essence, systemic and intended working... this truth and wisdom being a living source for ethics, ethical guidelines and pathways.

Chapter Five

Ethics, Ethical Guidelines

One of the truths we have come to understand is that if love is not present in the process, it will not be present in the outcomes.

If the ethics we generate, on any subject and for any process, are to encompass the whole of life - the whole of our living humanness - love must be present in the process. With love and through love it is possible to have and to hold as a serious intent this common aim for all ethical guidelines:

To advance humanness... to enrich life – the life of the whole, and the whole of life on this earth.

Reflecting on the "absence" of love we can see a possibility of enhancing existence – at least materially – but not of having access to the spirit and meaning that love brings.

Reflecting further we comfortably see that it is natural to want to move up planes of existence - to have a better life for ourselves, for our children, and for future generations. Love entering through the eye of the heart does not seek to deny or diminish this natural inclination seemingly inherent within us; rather love works to illuminate the truth and wisdom that would enable – that would let us see – ways of making progress that would advance our humanness, and would increasingly honor and harmonize with the whole of life.

With love in our process and holding in our hearts the common aim, "to advance humanness... to enrich life – the life of the whole, and the whole of life on this earth," the eye of the heart, reflecting on these five ethical guidelines, would seek to see their essentiality, wisdom, wholeness and systemic relatedness:

- No occlusion to the "tuning" of the planetary life energy field.
- On entering a territory, seek to understand and honor, rather than interfere with, the spiritual culture and other essential life processes.
- Honor and understand the working of the life of the whole.
- Pursue technology and approaches that enable community self determination and environmental amelioration.
- Create processes that engender understanding and discourage divisiveness.

Ethical Guideline One

"No occlusion to the 'tuning' of the planetary life energy field."

As we reflect on our experiencing - innerly experiencing - the planetary life energy fields of earth, we can with a bit of effort recreate – in the eye of our heart – the depth, the nature, and the significance of our tuning experience. We notice that as living human beings we can innerly organize ourselves in ways that we become truly receptive to that which is at work here. We can tune ourselves into the spirit and energy of the landscape before us... and more impressively this effort of ours brings about – within us – a tuning by the life energy field itself. The life energy field truly works on us, a phenomena recognized by the statements, "You go to the mountains to build your mind; you go to the ocean to soothe your soul."

Reflection on our experience of tuning into and being tuned by life energy fields leads to the understanding of the necessity for being-to-being interactions. Such interactions are not only energizing, but through patient receptivity on our part, bring about an access and elevation of spirit that itself leads to a coalescence – a oneness, a wholeness – not only within oneself, but with the whole of the universe. There is a deep sense of connectedness, at times a diminishment of self, that leads to a true appreciation – an overwhelming sense of awe and wonder - and in that moment we gain a taste of what it means to be fully alive – to be fully present to our living human nature.

Whatever is at work as we tune into and are tuned by life energy fields is undoubtedly significant and obviously an intended aspect of our design... and whereas we may be the highest order form of life on earth, we can readily witness and observe the tuning capacity of our fellow creatures of life – the flora and fauna of the life energy field. It too is not only an aspect of their design – their living nature – but an inherent process in the working of life itself, a process that produces a particularized living system relative to that planetary life energy field... a living system that is both produced and enhanced – in both vitality and viability – by particular structures.

Now we know from experience, that there are times when it is difficult, if not impossible, to innerly organize ourselves such that we can be receptive to tuning. Preoccupation of mind, hurriedness, inattentiveness, etc., all work to interfere with our tuning capacity. Likewise, what works innerly also works outerly. Structures and structuring can be present that interfere with our ability to tune into and be tuned by the life energy field. What is true for us is likely true for all of life's creatures.

Finally we understand that we are people of earth - people who live within and occupy particular life energy fields. Tuning into and being tuned by a life energy field gives us access to the virtue and spirit of the field itself... a process that, we are beginning to see and understand, enables us to come together in ways previously not attainable.*

* See "Becoming; Right for the Heart... Good for the Whole," Path of Potential, 2005, for understanding of what becomes possible when a community adopts the virtue of the land as their living philosophy.

Ethical Guideline Two

"On entering a territory, seek to understand and honor, rather than interfere with, the spiritual culture and other essential life processes."

Life's way is process... process that becomes more visible, more embraceable, through ebb and flow, cycles, cyclical patterns, and patterns within cycles. Culture is a process, a life process, an essential life process. It is the way of the people – the people's way of living, way of working, way of worshipping, honoring, dignifying, celebrating, etc. Culture, when it is truly working, is in reality a living being. As such, it experiences all that is true for life. It struggles to survive, to sustain itself, to develop, to evolve... it struggles to become.

Behind the visible manifestations of life lies the life force – that which continues to bring life into existence, and works to sustain life's course along the intended unfolding path. Once again, culture, like life, has that which lies behind its visible and embraceable elements - the systemic elements of values, rituals, totems, language, taboos, etc. Behind these visible elements of culture lies its essence, its virtue, its spirit – the particular manifestation of will entangled with love.

It is the active presence and manifestation of essence, virtue and spirit that are the ongoing source of what we have come to see as the spiritual culture of the people. Spiritual culture is the means by which the people can be and become that which they truly are… their way of being authentic throughout their journey. It is the way they stay connected to the Source of creation, their means of participating in the great unfolding, of taking on and fulfilling purposes… purposes that serve the whole of humanity, that work to advance our humanness along the intended path, and ultimately serve the whole of life. As such, interference, be it intentionally or unthinkingly caused, has consequences of significance to each and all – to each of humanity, and to all of life.

Life is about wholes; life always operates within and relative to particular wholes. These living wholes work to sustain their integrity by willfully sustaining their connection to their essence, their virtue, and their life purpose. It is within wholes that individuality is meaningfully manifested, and significance is realized.

Wholes provide for life a sort of boundedness, not a boundedness that would create a closed nature, but rather the nature of boundedness that brings about the orderliness and organizing necessary to carry out the intended work – the essential work, the purposeful work - of life within a particular whole. In the absence of the presence of such wholes, life cannot go about its work; life flounders, and access to the life force diminishes.

Seeking to honor the working of the life of the whole brings to the fore the notion of sustaining integrity - the necessity to ensure the actions taken do not violate or disenable wholeness and the capacity to bring about the orderliness and organizing required for life to carry out its work. This capacity has a component – a requirement – related to scale. Sufficient scale must be present to carry out the particular life's work. A rain forest in our back yard, for example, would not be of sufficient scale to carry out the life work of the rain forest.

A second consideration in regards to honoring the working of the life of the whole relates to structures and structuring. There are structures that can naturally integrate into the working of the processes of life, and there are those that cannot. For those that cannot, we either need to refrain from creating or using them, or we need to generate methodologies whereby they are prevented from entering into the working of life's processes. And in reality, even for those that can integrate naturally, there is most often, if not always, a need for thoughtful methodologies. With a bit of reflection we can see both the necessity for and the benefit of understanding the structuring that comes about as a result of the structures we create and introduce into living systems.

Honoring the working of the life of the whole also demands that we see and acknowledge the systemic nature of that which occurs within living wholes. Acting in accord with the reality that each element is related to and affects the working of the others is an inescapable requirement of honoring. "Touch one, you touch them all," may be a useful phrase to remind us – to awaken us – to the necessity of honoring the working of the life of the whole, and of the consequences should we willfully or unknowingly choose to disregard it.

Ethical Guideline Four
"Pursue technology and approaches that enable community self determination and environmental amelioration."

It is natural to want to move up planes of existence – to have a better life for ourselves, for our children, and for future generations. This natural inclination has been a significant force behind our use, pursuit and consumption of energy. In much the same pattern, this natural inclination is becoming a significant force for more and more people across the globe. The so-called thirst for energy is not diminishing, but increasing at significant rates, and is taking place in a world that exhibits characteristics of social/political instability and unpredictability. It is this uncertainty that has added a new dimension to the conversations and concerns regarding energy. "Securing energy" has been added to previous notions of "economics" and "environmentally friendly." This latest dimension has brought with it a sense of urgency seemingly greater than that which accompanied previous (but still valid) concerns.

And so, as is our pattern relative to matters of existence, we look to technology to provide solutions to the energy issues before us, solutions which if they are to be ethically guided cannot be pursued in isolation from, but of necessity within, the context of the whole – the whole of humanity, the whole of life. Grounded in the current situation (ours and the world's) as a reality before us, and holding the aim, "to advance humanness… to enrich life – the life of the whole, and the whole of life on earth," we can, and by necessity will go forth. With a bit of thought we can see clearly that "stopping" or "putting a halt to" current human activity (tempting as it may be sometimes) is not a viable option. We can, however, go forth in increasingly ethical ways.

Bringing ethics and being ethical relative to our development and use of technology represents for all of us a true choice, a turning point... one that demands understanding beyond that which can be achieved through argumentation... a turning point, an imposed requirement if you like, that is lifted up by the ethical guideline: *Pursue technology and approaches that enable community self determination and environmental amelioration.*

Community self determination offers the possibility to manage existence in a way that is open to and along the path of our potential, the path of intentionality, the path along which we can advance our humanness – move toward becoming fully and truly human. As humanness advances, peace – a particular manifestation of love at work within and among earth's people – becomes strengthened... and grows both in possibility and in realness of experience.

Environmental amelioration – bringing about an upward shift in our environmental state of being, an ongoing process of continuous improvement - takes us beyond just arresting entropy (run-down), and brings into play negentropy (run-up), which in turn results in actions that reflect a deeper, more wholistic, more systemic understanding... actions that reflect more consciousness, and more ableness for being conscientious. And as our work to enrich life - the life of the whole, and the whole of life - unfolds, harmony – the intended relationship between humankind, the community of life, the life processes of earth, and earth herself - also advances.

There is much hope in this turning point, this time of true choice, and in this shifting of patterns.

It is natural for a life species, a life community, a life system, to bring or impose particular demands – real requirements for continuing their existence and well being – on the processes of life. As human beings we have inherently within, a developable capacity to understand the life community members' requirements, as well as to understand that which is required to sustain the vitality and viability of the processes of life, a capacity that is being called upon out of necessity – true need – for ourselves, for our children, for future generations, for the whole of life. This capacity is the very capacity required of us if we are to carry out a stewardship role upon and for earth, a role which cannot be filled as individuals, but rather as community… a community that can generate images of the wholes involved, the right and good - right for humanity, good for the whole of life - working of each whole and the systemic nature of its working… a community that can create a process that generates understanding.

It is through understanding that true reconciliation can be achieved... a reconciliation that can elevate both human demands and life demands, two phenomena that often seem to oppose one another... a reconciliation that can bring into reality actions that simultaneously advance humanness and enrich life – the life of the whole, and the whole of life on earth.

This nature of understanding comes about through reflection and reflective dialogue. True reflection does not lend itself to (or in reality tolerate) the energies of argument, the energies that serve to divide. For the work of reflective processes is to gain understanding through the building of imagery – imagery coming from and working to create wholeness, a wholeness from which one can discover one's own role, one's work, the work of the heart.

And so what is seemingly being called for now is to take the energies of argument – the energies that are required to sustain the presence of a particular perspective or viewpoint – and transform them into willful actions of integration... willfully and reflectively ensuring that the particular perspective - its essence, its valuable contribution - is woven into the tapestry of the whole, thus enabling the community to generate and have access to wholeness and completeness of thought… and out from this living synergy of wisdom and reason, truth and knowledge, wise choices and conscious and conscientious decisions can be made.

Chapter Six
Community Principles

Communities, true communities, are in reality living wholes. As such they work to sustain their integrity by willfully sustaining their connectedness to their essence, their virtue, and their life purpose. Communities, as wholes, are places within which evolution – a process for creating a better world, for advancing humanness, and for enriching life - takes place.

Reflecting further on wholes, we can see that wholeness provides a path for will to enter into our processes such that our doing reflects what we are trying to be and become... will that manifests itself through spirit – through the elevation and evolution of human spirit... will that not only serves to advance our humanness, but also serves to increase our capacity to carry out our intended work... work and spirit deeply connected to virtue, essence and purpose.

Reflecting on our experience when wholeness is dissipating, we notice – we clearly see and feel – the diminishing of spirit, energy and aliveness. When our valuing for virtue and purpose diminishes, willfulness toward virtue and purpose becomes essentially nonexistent and lost, vitality begins to leave, and viability becomes uncertain and questioned… an uncertainty and tenuousness that allows that which is intended to be sourced in virtue (economics, legal-ness and exercising of rights, for example) to begin to operate as if it were the source… all of which tends to produce within, a growing sense of aimlessness - an increasingly real experience of having lost our way, of confusion over who we are and what we are trying to be and become.

It is a characteristic, an essence character, of our country – of America and her people - to be virtuous, to perform good deeds and to take on purposes that serve the larger whole – the whole of our country, the whole of the world. In the absence of this essence character, uncertainty, some confusion, and a real sense of stuckness seem to set in.

It is common for us as human beings to form communities – communities intended to work as wholes – in regards to fulfilling basic human and social needs, managing essential life processes, etc. Critical to the forming and sustaining of these is to have vivid imagery of the essential process of the particular community.

Reflecting on the essential process of the community of automobiling, for example, we can "see" the essence and virtue of *freedom*... a freedom of access... a freedom to seek out and be at one with others... a freedom to explore, grow and develop through real experiences; to move about, to seek out and develop opportunities, and to pursue one's potential. As our imagery develops, other wholes and processes begin to emerge: associated processes (such as flying, bicycling and walking) and related systems (living systems, health care systems, fueling systems and road systems, for example) affected by our current approaches and realities in regards to automobiling. We begin to see and experience the potential and power in seeing automobiling as a process versus limiting our scope to the automobile – a singular element within a systemic whole. And too we get a clear sense of what enables wholeness and what is contributing to dissipation of spirit, energy, and aliveness. Further imaging helps build an understanding that these dissipations are not burdensome problems, but rather sources of potential - the means for repotentializing, in an authentic American way, the whole of automobiling.

Principles, regardless of the nature of the community - the living whole of which we are members, or the living whole we are trying to create – are very helpful, and in reality essential... essential guidelines to not only keep us on the intended path, but also to keep us from unconsciously going off the path.

Principles are guidelines that describe our way of working – of being and behaving – as we go forward to engage community activity, pursue particular work, or take on particular roles. They work to help us take on the new pattern, the pattern made visible and chosen through reflection and dialogue. As such they serve to keep us on path. Also present in our set of community principles is a principle or two reflecting current experience – an awareness of an ingrained pattern, a pattern often so common in our current way of doing things, so comfortably familiar, so automatic that we may not immediately recognize that it is taking us off our intended path – along a different path... a path that most often reflects where we have been, rather than what we are trying to be and become. As such our principles commonly have a cautionary character such as "Don't add _____; don't seek _____; refrain from _____," etc. They serve to hold us on our intended path - keep us from spinning out, sliding off, or wandering – thereby sustaining our energy and spirit as we work to create a living community whole.

Chapter Seven

Scientific Theories & Knowledge

The advancement of scientific theories and knowledge has been instrumental in our ability to move up planes of existence. There is no doubt that science has been the primary instrument for the people in matters relating to existence… a reality perhaps not surprising since the primary domain of science is existence – in particular the physical, the material, the energy, and the how aspects of existence. We look to the ability of science to extend and exceed our human sensory perception to see what is at work in our universe, both above and below. We look to science to focus the "eye of reason" on subject matters of particular significance or consequence to ourselves and to life beyond ourselves. Through the eye of reason we hope and expect we will come to see and ferret out structures and structuring – how they work and come together – such that we can generate our own structures and structuring in intelligent and knowledgeable ways… ways that are both harmonious with our intent, and free of harm.

Given our historical pattern and the capability of science, it is quite natural that the people would call upon science to be a rightful and good instrument – right for humanity, good for the whole of life – relative to the critical subject of energy – securing energy. A call for science is in reality a call for the scientific community… a call for the particular community that is forming and needs to be formed around the subject matter of energy – securing energy.

This call by the people's community is a call that has a particular character and nature – an emerging clarity of hopes, expectations and requirements. We the people call upon our scientific community to engage in this scientific endeavor in concert with and with full regard for the aim and ethical guidelines expressed… and also to complement the wholeness imaged with completeness of thought, a completeness that no doubt will require a multitude of perspectives, perspectives that would not serve to divide, but rather work to create unity, even in the presence of competing thoughts.

This people's call upon the scientific community recognizes the reality of today… the expanding thirst for energy across the world, the unlikelihood of our willingly moving down planes of existence, and the necessity for wise restraint as we deal with this growing sense of urgency in regards to energy and securing energy… a reality accompanied by an expectation that progress will likely follow a path of "what to do right now," an understanding of ultimacy of energy source, and wise transitions that are both harmonious with and enabling of the ultimate source. This people's call also reflects an understanding of the resolving power of aim-directed, organized, scientific understanding, and along with that, the power – the willful power – that lies within the awakening of the virtue of America and her people.

Reflecting on what has been written here, we begin to see that what is being called for is reciprocal maintenance between the being of the people and the work and working of the scientific community. The people's aim of being more fully and truly human and becoming more able stewards of earth – life enrichers – looks to the science community to provide knowledge and intelligence such that how we manage our existence is harmonious with our aim… and the science community, as it engages in scientific endeavor, looks to the people for rightful support, real understanding, and guiding wisdom such that the possibilities it pursues enable us to continue to progress along the path of our potential… a reciprocal maintenance that calls for an interactive and understanding capacity between, within and among the people and the scientific community.

A note of hope – America's people, given an authentic understanding, will willfully bring about that which is required.

Chapter Eight
Community Dialogue

Community dialogue is truly working when wholeness is being imaged, is being generated, and is being experienced - when wholeness is alive and present in the process. The process itself begins to take on the character and role of being a guardian of wholeness. Reflective vigilance works to ensure the living presence of the community aim, a particular aim that has come into being through reflection, dialogue and contemplation... an aim that has a real sense of rightness and goodness about it... an aim that reflects and projects the essence of the community and what it is trying to be and become... an aim that is harmoniously guided by the virtue of the land upon and within which the community carries out its life processes.

The community aim, as its meaning deepens and as it becomes more integrated into the daily living and working of the people, becomes not only a source of spirit, but also a means for manifesting spirit – the spirit of the people... the spirit of the land. The character and expression of the community aim is a reflection of the commonly shared aim,
*To advance humanness… to enrich life -
the life of the whole, and the whole of life
on this earth,*
a reality that enables the community to experience a common bond, common work, and a relatedness with other communities whose aims are similarly sourced.

Into this dialoguing process - a process seeking to see wholeness, to see right and good working within the whole, and to see spirit, energy and love at work - enters a subject of significance to the community. And so the community dialogue takes on a particular orientation relevant to the subject. Imaging community dialogue with the intent of bringing life to the subject of affordable housing, for example, we can begin to see some of the images that would naturally show up… images that through continued dialogue and reflection would continue to be developed and enhanced.

Very likely a more wholistic image of affordable housing would begin to form. Affordability would begin to include consideration of energies - material energies, the life energies of earth, as well as the energies of well being - the energies that flow among and between people… people who share a common humanity, who share a desire for the

sense of belonging accessible within community... communities of particular nature and character – reflecting unique purposes, pursuits, and work... particular communities into which people – people from all walks of life – are drawn because for them it truly is a space where they have greater ableness to be... communities that have within, quiet spaces, sacred spaces, enlivening spaces, communing spaces, interactive spaces, etc. As the richness of life of true community begins to unfold, some previously held notions begin to fade or drop off. We notice "low income" is really neither a useful nor a valid starting point for wholeness - wholeness of the individual, wholeness of the community... a realization that becomes a source of creativity – creative approaches beyond the current, the common – for the financing of these affordable communities.

With wholeness of image present within community members, interactions with scientific community members can fruitfully go forward. Completeness of thought, knowledge and know how – the capacity to soundly generate required structures and structuring – can be applied to bring into existence that which intuition has wholistically conceived. Thus, as people with understanding of engineering, carpentry, masonry, design, architecture, etc., come together with various community members, homeowners, etc., the community aim and intent is neither discarded nor ignored, but rather becomes more real, more alive... and the working of reason - its gift, its role, its potential - is increasingly understood and appreciated by all.

Chapter Nine

Intentional Manifestation

Intentional manifestation is about choice – conscious choice, conscientious choice... an ongoing process of choosing. Going forth from reflecting and dialoguing, even with some clarity regarding some aspect of work or role that we might take on relative to the larger whole, we experience and we become subject to the temptation (if not the tendency) to not draw upon will, but rather surrender to the comfortable, the familiar, the knowingly doable, the acceptable... and we may even discover ourselves developing rationalizations for our choice.

Having true intent, we gain access to conscience – to seeing more and more clearly the right and the good. With intent, true intent, we are willfully stepping into the stream of evolution, the stream of the intended unfolding, and perhaps not surprisingly, we begin to experience an unfolding – an increasing clarity and a deepening sense of rightness of path, that this is a path of potential: essence becomes more accessible and present; we grow in certainty about the work of our heart; our work and being begin to reciprocally maintain one another.

With intent, purpose becomes necessary and real. Purpose enables us to sustain intent… and purpose itself deepens in meaning. Our sense of purpose, that we have purpose, a reason for being – roles to play, work to carry out, work of significance regardless of scale – lies right before us. Reflecting further on intent, we notice it too deepens through conscious and conscientious effort. It becomes more and more an integral part of ourselves. What may have begun as a declaration – a speaking out loud of what has entered within – soon follows a deliberate path, a path of many choices, a path of upward progressions, a path that intersects and engages increasingly larger wholes.

It is the points of choice that provide opportunity and means for will to enter, not will in the sense of strong determination – the will we often experience as our own - but rather will that enters into and through being... will that we, through our instrumentality – our purpose, the work of our heart, the clarity of service to the larger whole – draw within and through ourselves... a process by which will can enter into the working of the world. True choice requires both the presence of consciousness and conscience – the seeing of the whole and clarity of rightness and goodness. The process of consciously and conscientiously choosing – through repeated cycles of choice – is the means by which will is strengthened. As we "exercise" will it becomes more integral, more accessible and more present within us.

Principles, like purpose, become essential as we actively walk along our path of intentional manifestation. We know from experience that establishing new patterns while resisting the temptation of old patterns requires sincere effort on our part. This effort is greatly enabled through principles... principles that are shared – commonly held – by the community. Principles provide rightful boundedness... real means for sustaining our intended direction and path, as well as "preventing" us from going astray, either habitually or unconsciously. Principles serve as consciousness and conscientiousness maintainers. They helpfully work to ensure that which is manifested through our efforts is in concert with intention – the intended unfolding.

Intention is firmly rooted in essence and virtue, and it reaches towards purpose – purpose that enables the realization of our aims... aims that are harmonious with the commonly held aim, the aim that reflects the intended unfolding and the work of the heart of humanity:

To advance humanness... to enrich life –
the life of the whole, and the whole of life
on this earth...

an aim through which love enters into the working of the world... love that makes possible the realization of our potential and our participating in the intended unfolding.

Chapter Ten

Upward Progressions

Living experience and the experience of living from intent and in concert with intent take on the way of unfolding upward progressions... progressions that have as their source the growing realization of and the embracement of the truth that we are not the source, rather we are instruments, instruments that have purpose, work and roles to fill... progressions that reflect and embody the aim of humankind,

> *To advance humanness... to enrich life –*
> *the life of the whole, and the whole of life*
> *on this earth...*

the aim we are intended to share... an aim we have a genuine birthright in... the aim that represents the work of our heart... the aim through which we as humanity can "be true to thyself"... the aim that reflects our reason for being – our reason for being here. Earth has a reason for being, and she works diligently and tirelessly to fulfill it... and we, as people of earth, by saying "yes" to intent, "yes" to the common aim of humankind, willfully choose to work and walk along the path of becoming fully and truly human – a path congruent with the working of earth.

Faith, which at first seems more like a "leap of faith," deepens and evolves. Our progressions bring about a growing faith and confidence in intent and design, the intent and yet to be realized potential that lie within and behind the image of humankind – the image of humankind held by the Source of creation… a faith that is further strengthened by our witnessing of what unfolds as will and love enter into the working of our community… the progression of inclusivity - the progression of our capacity, through sharing of intent (the intent of the land, the aim of our community, and the shared aim of humanity) to come together and transcend that which would normally divide, that which seemingly from habit - ingrained habit - works to divide, to disrupt the creation of wholeness and the path toward oneness.

We see further progression within ourselves and our gatherings. The inner processes of reflecting and dialoguing become more common, more integrated into our ways - the way of the community. Where once there may have been a single gathering, we find a growing number of intuitively led gatherings... gatherings that - through harmonizing with the virtue of the land, having shared community aim, and the emerging living philosophy of the community - work to bring out an increasing sense of unity within the community. More and more, community issues are addressed from the perspective of potential versus the perspective of problem.

Life subjects as well as issues begin to be reflected upon from the perspective of potential. In depth reflection and dialogue are pursued and sustained until sufficient wholeness and wisdom are present to provide guidance... ethical guidance... practice-able and practical guidance... guidance enabled by principles. Subjects that represent life cycles from birth and birthing to death and dying, when viewed from the perspective of potential, are seen in a new light, a more meaningful light, a light rich with hope. Seeing the essence, the work, and the purpose of the stages of life enables the community to organize itself for successful completion and development of the stages, all of which contribute to the ongoing development and enrichment of a philosophy of life, a community philosophy of life.

Life itself takes on new, deeper, richer meaning... a greater sense of purpose... more lived from the spirit and increasingly a manifestation of spirit. Humankind steadily gains more access to awe... the great awe of creation... the wondrous manifestation and working of life on this planet earth.

Chapter Eleven

Community Integration

Community integration is about soul... soul building... the building of the soul of humanity... about joining in the work of building the soul of humanity. It is about the work, the work that is the source of a stirring, an awakening, that is being experienced by more and more people... people who intuitively know - an intuition increasingly factually supported - that the path we are on is not a path of potential, not a path leading to advancing humanness, to enriching life... people who in their heart of hearts are living in the question - a persistent questioning - "There must be a better way?" ... there must be a way of hope, a way worthy of faith - of having faith and of placing faith in.

Community integration is working, is at work, when spirit is entering in through essence and virtue, and is being manifested through structures and structuring that have the character of intentionality... structures and structuring that are not merely manifestations of "what's possible," but rather those that reflect our intention, our shared intention, to advance humanness, to enrich life... structures and structuring that are along the path of evolution - the path of our becoming fully and truly human.

Community integration understands fully and truly that if love is not present in the process, it will not be present in the outcomes... and that it is through process that love, spirit and will enter into the working of the world, into the ways and working of life – human life, the whole of life. And the work of community integration recognizes that in a very real sort of way, evil is not so much about presence as it is about absence – absence of good... good which truly has one Source, the one and the same Source from which love emanates. It is through love that good enters into the world.

Reflecting on the "Cycle of Understanding Relative to Life on Earth" (p.19), we notice the likelihood of many cycles, and of cycles within cycles. And too we realize that understanding generated at one point will naturally elevate and deepen the understanding at other points... an experience and manifestation that will enable our continued progression toward our aim. We can see that with shared intent and a common aim, the understanding and wholeness generated through reflection and dialogue within one gathering can readily be integrated into the forming and working of our community – the larger whole within which we live and work. This working pattern can readily expand beyond a particular community to provide the means for integration – soul building work - among communities, which in turn further develops the understanding and wholeness that can be integrated into the forming and working of the community of humanity... all of which is possible because our community's aim is congruent with the aim, the work of the heart of our fellow communities, which in turn is harmonious with the work of the heart of humanity... our humanly shared aim,

To advance humanness... to enrich life -
the life of the whole, and the whole of life
on this earth,

the intended aim for humanity.

And so, the eye of the heart can see that with diligence and deliberation, willful attention and intention, the path of our potential can be walked upon and unfolded before us... a path made visible by bringing voice to intuition, and courage to conscience... a path of faith, of faith in action... a path which lifts up for us a possibility – a real possibility – of our, the whole of humanity, becoming a unified compassionate family... an intended becoming.

A Final Reflection...

GATHERING OUR COMMUNITY

Perhaps now is a good time to gather our community – our community of friends, family, colleagues, etc., gathering together to read aloud, reflect and dialogue.

... And as we gather our community, let our spirits be lifted and our hearts filled with hope... and joy... brought about by this truth:

*We do not have to reinvent the human being.
All that is called for, all that is needed
is fully present within our design.
In reality, all that is being asked for is that
we live it out as intended.*